SMART MANUFACTURING
TERMS YOU NEED TO KNOW!

MORE THAN 100 TERMS
EXPLAINED FROM
3D PRINTING TO ZIGBEE

Text and Illustrations by Mike Nager

Published by Industrial Insights LLC New Jersey
www.industrialinsightsllc.com

Smart Manufacturing Terms You Need to Know!

More than 130 Term Explained from 3D Printing to Zigbee

ISBN-13: 978-1-7363625-2-5 E-Book

ISBN-13: 978-7363625-3-2 Paperback

Our publications are available at special quantity discounts for use in sales promotions or educationaltraining programs

To join our mailing list and become eligible to receive a free chapter of a future

publication, register at

www industrialinsightsLLC com

To shoot us a message or interesting story related to Manufacturing, send it to:

SmartStudents@industrialinsightsLLC com

Twitter: @Ind_Insights

LinkedIn: https://www.linkedin.com/company/industrial-insights-llc/

Contents

Preface

The year 2021 is turning out to be a pivotable one in the world of manufacturing!

As I write these lines in March 2021, the Covid 19 vaccinations are rolling out across the United States. A surge of optimism is taking hold that life will return to "normal" within the following months.

But global supply chains appear more fragile than ever before. This month, a towering cargo ship, as long as the Empire State Building is high blocks the Suez Canal, stopping traffic in both directions. Estimates are that the cost of interrupted trade is $10 billion per day.

A worldwide shortage of semiconductor chips is idling automotive manufacturers everywhere – production has ground to a stop.

These recent events are accelerating the re-shoring movement to locate manufacturing close to its customers. Smart Manufacturing, sometimes called Industry 4.0, promises customized products created with less waste. New business models and new smart products further change and create opportunities for those with bold visions.

This compendium will serve as a reference to Smart Manufacturing's terminology.

Here you will find many terms that predate the concept of Industry 4.0 but are still crucial to enabling it -- and illustrate how progress builds upon past achievements. And here you will find at least a few new words and concepts as the Internet of Things and Information Technologies transform manufacturing.

Acknowledgments

For my colleagues, customers, clients that I have worked with over many years, who helped me understand the world of Smart Manufacturing and Industry 4.0.

Specials thanks to Garrett Schmidt and Christine Frank who helped provide content in this publication.

Disclaimers

The following disclaimer applies to any content in this book:

This book is commentary intended for general information use only and is not investment advice. Mike Nager and/or Industrial Insights LLC do not make recommendations on any specific or general investments, investment types, asset classes, non-regulated markets, specific equities, bonds, or other investment vehicles. Mike Nager and/or Industrial Insights LLC does not guarantee the completeness or accuracy of analyses and statements in this book, nor does Mike Nager and/or Industrial Insights LLC assume any liability for any losses that may result from the reliance by any person or entity on this information. Opinions, forecasts, and information are subject to change without notice. This book does not represent a solicitation or offer of financial or advisory services or products. This book is only market commentary intended and written for general information use only. This book does not constitute investment advice. All links were correct and active at the time this book was published. All content is the personal opinion of Mike Nager and does not necessarily represent any view or position held by any past or current client or employer.

Designations

Designations used by companies to distinguish their products are often claimed as trademarks. All brand names and product names used in this book and on its cover are trade names, service marks, trademarks, and registered trademarks of their respective owners. The publishers and the book are not associated with any product or vendor mentioned in this book. None of the companies referenced within the book have endorsed the book.

3D Printing

3D printing is a slang term for a sub-class of Additive Manufacturing. It allows one-off or small batch production of individual parts. 3D printers can be leveraged into the manufacturing process of Smart Factories, providing flexible production of small orders. Generally, a 3D computer file describes the object so the machine can then make the part.

The technology is advancing rapidly in speed and the types of suitable materials for the process. At first, the printers could only use various types of plastic, but now ceramic and metal printers are available.

The technology's first industrial uses were to create models or physical prototypes only, not parts used in the actual application. Today's 3D printers are capable of manufacturing components used in many different industries and applications.

3D printing also allows for creating complex shapes that would be nearly impossible to do with Subtractive Manufacturing techniques. See Additive Manufacturing and a graphic of the leading additive manufacturing technologies in Appendix II.

5th Generation or 5G Network

5th Generation cellular mobile networks or 5G wireless systems are the next telecommunications standards that will provide increased bandwidth, higher download speeds, and low latency. The 3GPP (3rd Generation Partnership Project) specification group ensures compatibility among different manufacturers and providers.

The 5G specification includes several different spectrum bands, including 4G LTE, Sub-6GHz, and mmWave. Download speeds in the Sub-6GHz and mmWave bands will reach up to 1.5Gbps with ultra-low latency suitable for real-time control.

Since many IoT devices will connect to the network, it will support 1 million devices per square kilometer or nearly 2.6 million devices per square mile. It could replace Wi-Fi and other wireless networks.

Applications include improved home and business broadband access, autonomous automobiles, infrastructure, IoT, and IIoT devices.

5S

5S applies to workplace organization and uses a list of five Japanese words to organize the theory – seiri (sort), seiton (set in order), seiso (shine), seiketsu (standardize), and shitsuke (sustain).

5S creates an efficient, organized, safe, and clean workplace. Additionally, it builds a culture of standardization for how a company, a division, or a process should operate and enables "Just in Time" manufacturing.

Sort: Removing all unnecessary items in each location

Set in Order: Placing necessary items in the optimum location

Shine: Inspecting and cleaning area and tools regularly

Standardize: Establish procedures ensuring that the first three methods(Sort, Set, Shine) are consistently applied.

Sustain: Empowers workers to "do" without being explicit in instructionsfrom management, to sustain the 5S approach

Additive Manufacturing

Additive Manufacturing refers to a technology that builds objects by adding layers of material instead of removing excess materials. A Computer-Aided Design (CAD) drawing describes the item in detail so the machine can manufacture it.

Types of Additive Manufacturing include 3D Printing, layered manufacturing, additive fabrication, Rapid Prototyping (RP), and Direct Digital Manufacturing (DDM). Nearly every industry uses Additive Manufacturing for small production runs.

It has extensive applications in the industrial world and represents a revolution in manufacturing processes and systems. See graphic of the primary additive manufacturing technologies in Appendix II. Below is a photo of a 3D printer.

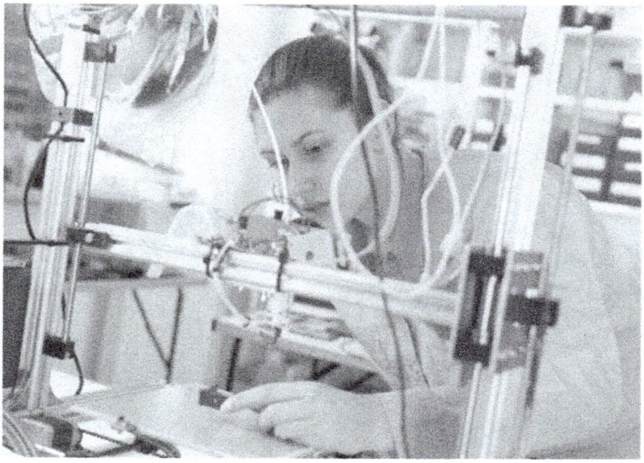

Advanced Manufacturing

Advanced Manufacturing is one of many terms that describe the transition from manual manufacturing operations to the use of machinerycontrolled by a computer. Advanced Manufacturing refers mainly to the use of CNC machine tool technology, especially in the United States. Using machines improves output, product quality, and competitiveness in the manufacturing sector.

The title overlaps in meaning with Smart Manufacturing, Industry 4.0, and the Industrial Internet of Things. Addendum I shows a graphic of the different terms. See Computer Numerical Control (CNC) and the machine tool below.

Advanced Message Queuing Protocol (AMQP)

AMQP is an open standard for message-oriented middleware that allows information generated by different vendors to communicate. The OASIS industry standards group governs it. AMQP originated from the financial industry (JP Morgan) in 2003. Similar technologies are MQTT and JSON.

App Store

The App Store is an online storefront that provides application software products, known as "Apps," to be purchased and downloaded. App stores are available for smartphones, tablets, or computers, and industrial equipment.

Concerning Industry 4.0, the immediate availability of new capabilities creates flexibility and versatility for robots and other devices, such as 3D printers, automated guided vehicles, smart warehouses, and more.

The manufacturers of this equipment can sell a base model for a lower price. If the customer needs more capability, they purchase an "unlock" code to enable those additional features.

Some industrial apps include data sharing, process control tuning wizard, maintenance scheduling, alert and alarm dashboards, valve movement, instrumentation loop drawings, PID (proportional, integral, derivative) loop simulator,

5S-Lean Audits, Energy Efficiency Checklist, Food Safety Compliance Checklist, Warehouse Management,

Application Architecture

It consists of a set of application guidelines and architectural components and may include messaging standards for accessing applications. This standardization is expected to enhance the organization's ability to extract data from applications (and hence, value), link disparate business and manufacturing processes, and ultimately eliminate redundancy between applications across the Industry 4.0 enterprise.

Application Programming Interface (API)

An API is a way for an application to present itself to other applications to share data. An API defines the types of requests that can be made to the application, how to make them, the conventions, and data formats to use, and more.

There are APIs for programming languages, software libraries, operating systems, and software applications.

Monitoring and Reporting APIs communicate information about how manufacturing equipment is operating to optimize performance. Visibility into the machine's productivity, material costs, and maintenance history provide valuable insights into the operation. APIs are also used to provide real-time status data and alerts to operators.

Command and Control APIs provide a method of remotely starting manufacturing operations from anywhere in the world. When combined with technologies such as 3D Printing, which can be deployed globally – or even inside a customer's facility – the time and logistical cost savings can be significant.

APIs also communicate information outside the company to suppliers or customers, playing a part in XaaS.

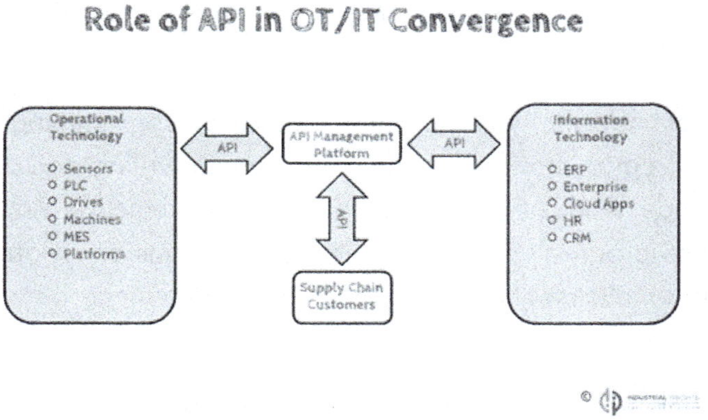

Artificial Intelligence (AI)

Artificial Intelligence is the development of computer systems able to perform tasks usually requiring human intelligence, such as visual perception, speech recognition, decision-making, and translation between languages, and to take actions that maximize its chance of success at some goal.

Artificial Intelligence is exhibited by computers that have "learned," based on processing large amounts of data, including historical facts and information on their environment.

In the Industry 4.0 context, environmental information comes from sensors and other inputs. Pattern recognition is applied to historical data by AI to make predictions and optimize scenarios to maximize results.

Much AI work is centered around inspection systems using cameras for quality control applications. State-of-the-art systems can analyze as few as five images to determine imperfections in under a second.

Other industrial AI applications include programming robots from a drawing instead of a human writing each line of code. And AI is being used to power Predictive Maintenance programs by sorting through data sets and warning of anomalies that may indicate failure is imminent.

Other applications include machine learning algorithms, motion sensors, and machine vision to spot and even predict defects. See Machine Learning, Deep Learning, and Addendum IV.

Asset Administration Shell (AAS)

A popular rising Industry 4.0 term, the administrative shell, describes integrating real-world assets into the world of information. Assets that integrate into Industry 4.0 include machines, raw materials, parts, drawings, schematics, wiring diagrams, contracts, orders, and many otheritems.

AAS is a virtual representation of a physical Industry 4.0 component using a standardized form describing the asset (a part, component, or machine) in a digital file. The digital file becomes a key part of creating a Digital Twin.

The AAS integrates the asset into the Industry 4.0 communication framework by making it addressable and precisely identified. The AAS covers the complete lifecycle, from engineering, purchasing, commissioning, maintenance, and end-of-life.

The AAS of a component knows the proper properties and capabilities, knows where to find the information about the component, has a specific standardized syntax, and can be realized as a static information source, like an HTML file, on a field device, in an IoT gateway, or the cloud.

Attack Surface and Vector

The attack surface is those elements of a system or equipment vulnerable to attack from a cybersecurity perspective. In general, manufacturers' attack surfaces are becoming more numerous as more mobile devices are used by customers and employees, as work-from-home becomes more normalized.

The attack vector is the actual malware (e.g., viruses, e-mail attachments, web pages, etc.) by which an attacker gains access to an entity or the ability to cause harm.

Automated Guided Vehicle (AGV)

AGV is a driver-less guided transportation system that follows markers, such as magnetic strips, bar codes, or vision systems that determine the paths an AGV needs to follow a specific route.

Manufacturing, warehouses, and other logistics companies use AGVs for material movement by transporting pallets, cartons, totes, or individual products. In some cases, AGVs replace forklifts; in others, they replace conveyors.

AGVs are incorporated into manufacturers' software platforms, especially ERP and MES suites, providing visibility and product flow traceability.

As AGV technology continues to advance, the definition blurs between them and mobile robotics. Mobile robots can follow a free-form path rather than a rigid pre-defined one, but AGVs are also advancing in this area.

Automation Markup Language (Automation ML)

Automation ML is a common data format that provides interoperability between digital tools used in production and planning. It is based on XML for the storage and exchange of plant information. The goal is to allow communication between different plant viewpoints, such as electrical, mechanical, HMI, PLC, and robot control. It was initiated in 2006 by the companies ABB, KUKA, Rockwell Automation, and Siemens.

Augmented Reality (AR)

Augmented Reality (AR) is an interactive experience of a real-world environment, where the objects in the real world are enhanced by computer-generated perceptual information, sometimes across multiple senses.

This technology adds digital information to existing environmental information, displaying them together to provide a composite picture that is much more informative. Unlike Virtual Reality, Augmented Reality does not create an artificial environment but overlays the new information onto the real one. In the Smart Factory, Augmented Reality assists in assembly, machine operation, maintenance, and training.

A key function in Augmented Reality uses the cameras that are integrated into a wide range of mobile devices to give the user a combination view ofthe real world and multimedia contents. The user perceives Augmented Reality via an optical head-mounted display, a smartphone monitor, or gloves that interface with other human senses.

Availability and Availability Loss

Availability is the amount of uptime that a machine runs and produces a product and is a percentage of the total planned production time. Availability is one of the three factors of Overall Equipment Effectiveness or OEE.

Availability Loss includes events that stop the planned production and include unplanned equipment failure (what other kind is there, though, really?), or material shortage, or planned changeover time.

AWS

Amazon Web Services is a scalable, secure cloud platform that provides services, applications, and database functionality to users. Users can pull Dashboard information without needing to store any data locally. AWS is an OPC Foundation member.

Azure

Microsoft Azure is a scalable, secure cloud platform that provides services, applications, and database functionality to users. Users can pull Dashboard information from Azure without needing to store any data locally.

Batch Size One (Lot Size One)

Batch Size One refers to one of the key benefits of Industry 4.0: creating personalized products. Producing customized products, one at a time, differs significantly from the mass production model used in Industry 3.0 operations.

Cost-effectively creating short product runs requires a great deal of automation and flexibility in the manufacturing environment. The increased costs of this complexity are offset by reducing unskilled and semi-skilled labor costs and, in many cases, lower logistics costs.

Creating personalized products close to the end-user provides a business incentive for manufacturers to locate their operations in developed countries, that previously relied on offshoring manufacturing, to low labor cost countries. This movement of manufacturing back to the US, Canada, and Europe is known as Re-Shoring.

Big Data and Analytics

Industry 4.0 operations rely on digitally connected devices, with a large volume of data being recorded, processed, and stored at any point in time. Big Data Analytics has enabled the Smart Factory to examine these data sets to make associations and spot patterns and trends.

The number of insights analyzing the data and producing actionable information spans the entire enterprise and includes production scheduling, costing, planning, Predictive Maintenance, and theprevention of bottlenecks.

Analyzing correlations present in Structured Data and Unstructured Data sets allows the manufacturer to uncover opportunities for cost-saving. See Data Analytics.

Blockchain

A blockchain is a continuously growing list of decentralized, digitized public records, called data blocks, of all cryptocurrency transactions linked and secured using cryptography.

The blockchain principle takes up the growing trend of decentralized dataadministration and processing. It is a chain of data blocks in which transactions are linked and examined.

Blockchain technology may be one of the most disruptive innovations of the 21st century. Technology, although being complicated to understand, represents a transformation in the way we currently conduct transactions.

Blockchain technology can record any digital information: birth certificates, title deeds, votes, financial accounts, product data, formulas,contracts, and more. All this is performed more quickly, securely, and transparently than with traditional alternatives.

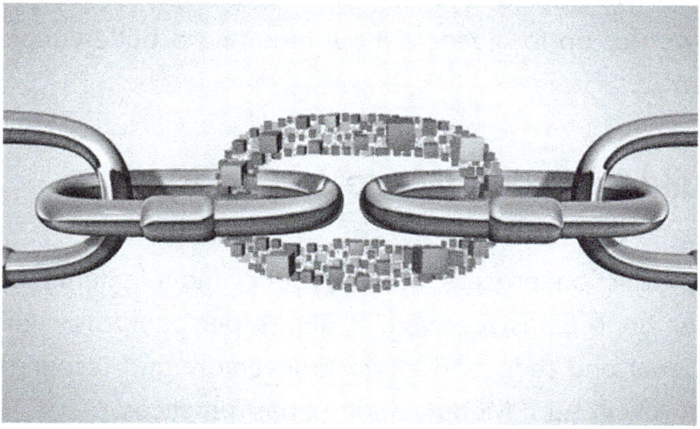

Bluetooth

Bluetooth is, traditionally, a short-range, wireless technology standardized by IEEE 802.15.1 and managed by the Bluetooth Special Interest Group.

Bluetooth Low Energy BLE (Bluetooth 4.0) is a lower-energy-consumption version of the Bluetooth wireless communications standard, which constantly runs, announcing a device's presence to local sensors, optimizing battery life for the device in question. In the IoT, BLE allows for precise location and feature tracking without reduced battery life.

Advances in the Bluetooth standard are increasing the performance in several areas. Bluetooth 5.0 allows operation over greater distances, up to a mile away. Bluetooth Mesh creates a Many-to-Many network instead of a One-to-One.

A direction-finding function of Bluetooth allows for exact indoor location and asset tracking of equipment. Bluetooth communication is relatively secure, and devices can self-power for up to a year without requiring a battery replacement.

Build to Order (Made to Order)

A production process where product manufacturing starts after an order is received. It allows personalization of the product and reduction of waste inventory and resources. It aligns with Lean Manufacturing's best practices.

The concept of Industry 4.0's Batch Size One is a specific type of Build to Order process and represents a "pull" production method where customer demand triggers the manufacturing process. The pull production process differs substantially from the "push" production method. Mass production "Make to Stock" operations try to "push" products out to the sales and distribution channels to find paying customers, hopefully.

Financial benefits include reducing inventory costs, specifically Work in Process (WIP) and Finished Goods inventory. In general, the positive monetary impact of not having to store WIP and Finished Goods is immense.

Business Intelligence

A set of methodologies and tools that analyze, report, manage, and deliver information relevant to the business, includes dashboards and query/ reporting tools similar to those found in analytics.

One key difference between analytics and BI is that analytics uses statistical and mathematical data analysis that predicts future outcomes for situations. In contrast, BI also analyzes historical data to provide insights and trends information.

Chaku Chaku

A Japanese term meaning "Load Load" describes a type of Manufacturing Work cell used in semi-automated lines. Semi-automated lines attempt to seek the proper balance between automated and manual production.

In Chaku Chaku, a human operator loads a machine with parts. This procedure's automation tends to be complex and expensive, so using a Chaku Chaku line can be appropriate for many applications.

The machine performs its application and typically ejects the part, which is relatively simple to automate. The human worker then takes that part and loads it into the next machine.

The link shows a Chaku Chaku line:
https://youtu.be/5iNXiOCY2HY

Cloud Computing

Cloud Computing enables the Industry 4.0 environment to be fully connected, using the Internet, data security, and storage facilities. This remote centralization of business information provides a computing platform for Digital Manufacturing without purchasing hardware and software. See Software as a Service.

Cloud computing uses a network of remote servers to host your data and applications on the Internet, not on a local PC or server. It allows infinite space to store data and eliminates data loss potential since cloud computing vendors create backups. It also enables collaboration, irrespective of geographical location, and offers always-on availability.

Some examples of cloud-based platforms used in Smart Manufacturing include Azure, HANA, Siemens MindSphere, Rockwell FactoryTalk, Amazon Web Services,

Cloud Robotics

Cloud Robotics is a term created by James Kuffner in 2010. He was a Google employee at the time. The idea was to provide an extended brain for robots.

Communication between the physical and digital worlds is controlled in the cloud and extended to robots used in mobile applications. In Smart Manufacturing, robots access decentralized data in networks or the Internet, boosting their performance and flexibility.

Cobots (Collaborative Robotics)

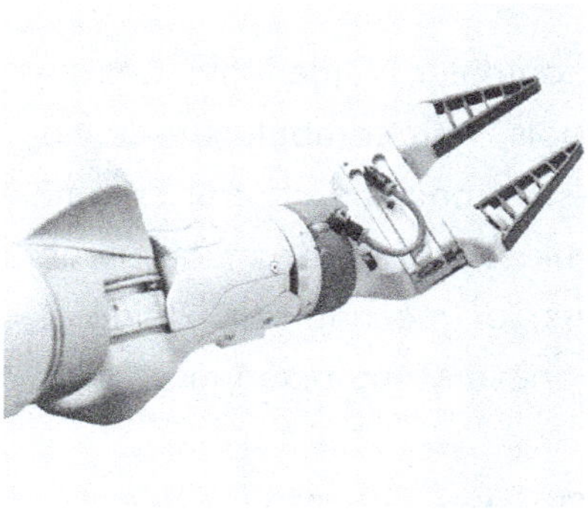

A cobot, or collaborative robot, is a robot intended to interact with humans in a shared workspace.

Cobots provide a "third arm" to human workers and is safer by design because they don't work in cages as industrial robots do. Typically, they will not have any hard, sharp edges and are constructed of relatively lightweight components.

The maximum speed of the cobot may be limited. Force sensors provide feedback to the robot, so if an unexpected object contacts it, the cobot comes to a stop. A human can push a cobot out of the way during a power failure because it doesn't lock in place like earlier robot designs.

As a new technology, manufacturers of cobots have created more intuitive programming techniques that are easier and faster to learn. Several manufacturers allow the robot to be programmed by manually pushing it through its routine, which is then recorded and duplicated. In parallel, manufacturers of end-of-arm tools, like grippers, create safer and more flexible designs than were previously available, by avoidingsharp edges and using softer materials.

Computer Numerical Control (CNC)

Computer Numerical Control is the automation of machine tools, with computers executing pre-programmed sequences of machine control commands. This contrasts with machines that are manually controlled by hand wheels or levers, or mechanically automated by cams alone. See Advanced Manufacturing.

Customer Journey

The term Customer Journey designates the data-based analysis of buying decision processes. If the customer purchases online, it is possible to reconstruct the client's decision-making path using special tracking tools. For example, when using the tool, it is possible to determine how many advertising contacts were necessary to activate a specific product's purchase.

Cyber-Physical Systems (CPS)

CPS are objects with embedded software and electronics connected in a system, such as robots, drones, and other movable machines. This way, physical and mechanical objects, and processes are associated with software-controlled objects and processes – with the real and virtual worlds converging. CPS can be used for traffic control or for managing intelligent electricity networks.

A system that links real (physical) objects and processes with information-processing (virtual) objects and processes via open, in some cases global, and constantly interconnected information networks. Note: A CPS optionally uses services available locally or remotely, has human-machine interfaces, and offers a dynamic adaptation of the runtime system.

Cyber-Physical Production Systems (CPPS)

CPPS is the name for the connected machines in a Smart Factory, where one piece of equipment's status and actions affect the others. Sensors are used for the data collection on each device analyzed and used to provide information on the overall production system's performance and condition.

Physical and digital items are connected, monitored, and managed with computer programming and algorithms.

In intelligent production, the CPPS unit controls itself. It makes decisions based on individual parameters, for example, deciding "does the relevant function/capacity exist for the product's requested version?"

Cyber Security

In a digitized environment, protecting important company information becomes increasingly important. Cybersecurity means all the technologies and services that protect the company from any digital attack or data loss. Potential threats from inside and outside the organization need to be identified, and policies implemented to prevent successful attacks.

Within a manufacturing operation, Cyber Security protects production equipment that uses Industrial Ethernet and has access to the Internet. See Attack Surface and Vector.

Cycle Time

Cycle Time is how long it takes to complete a task and is an essential manufacturing Key Performance Indicator or KPI. For example, it is used by ERP and MES systems for scheduling, purchasing, and determining production costs. It is also a key component of calculating Overall Equipment Effectiveness (OEE).

Dashboard

Displays graphical information about the IIoT ecosystem to users and enables them to control their machine, device, or factory. The image is displayed on a Human-Machine Interface (HMI), personal computer, or a mobile device.

Dashboard graphics show gauges and charts, providing accessible information to the human observer. A heat map provides an overview of the errors or other events over time, allowing the operator to make a quick analysis. Widgets are boxes within the dashboard that display data in a bar chart or some other visual representation.

Data Architecture

The data management architecture refers to the data models created within the enterprise to facilitate the creation of a standard view of the data resources within the organization and common methods for assessing and integrating data, including augmented intelligence.

This standardization allows all users better access to business data, integrated business views dependent upon integrated data, and data- facilitated (analytics) real-time response to market opportunities.

Data Attributes

Variety refers to the different types of data we can now use. At least 80% of the world's data is unstructured and can't easily be put into tables. With Big Data technology, we can now better analyze unstructured data, including messages, photos, sensor data, video, or voice recordings, and bring them together with more traditional, structured data.

Velocity is the speed at which new data is generated and the speed at which information moves around. Big data technology allows us to analyze the data while it is being created, without ever putting it into databases.

Veracity refers to the messiness or trustworthiness of the data. With many forms of big data, quality and accuracy are less controllable. Volume often makes up for the lack of quality or accuracy

Volume refers to the vast amounts of data generated every second, which increasingly makes data sets too large to store and analyze, using traditional database technology. We can now store and use these data sets with distributed systems, where parts of the data are stored indifferent locations and brought together by software.

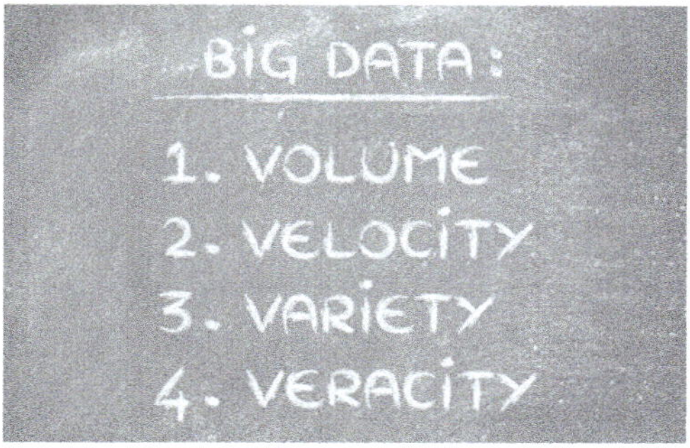

Deep Learning

Deep learning is a subset of Machine Learning and Artificial Intelligence methods, where the system learns by analyzing large data sets and making predictions that become more accurate over time. Deep Learning can analyze Unstructured Data in contrast to other Machine Learning techniques that rely on Structured Data sets. Deep Learning doesn't necessarily require supervised learning, where a human provides input but can detect unlabeled data patterns. See Addendum IV.

Defense Innovation Unit

The Defense Innovation Unit is a department of the United States Department of Defense that aims to scale technology at commercial speeds for national defense. It focuses on advanced energy and materials, Artificial Intelligence, autonomy, cyber, human systems, and space.

Delayed Differentiation

Delayed differentiation is a technique to delay the final production of a product until the very end so that customization can occur. It is a way to achieve Batch Size One production.

Department of Defense Architecture Framework (DoDAF)

The standard used in the United States defense industry for organizing complex systems-of-systems.

Descriptive Analytics

Descriptive analytics creates a summary of historical data to produce useful information or to answer the question, "What happened?"

Digital Enterprise

A Digital Enterprise extends Industry 4.0 outside the factory walls. A connected supply and value chain allowing greater visibility of locations and activities across the whole enterprise, rather than just the internal processes. It is tightly coupled with the concept of the Digital Supply Network.

Digital Shadow

Digital Shadow is a virtual image of a real object where there is a one-way flow of information from the physical to the digital object. A change in the physical object results in a digital object change as well, but not the reverse. See Digital Twin for a closely related manufacturing topic.

Another definition of Digital Shadow is related to the information that is left behind when a digital service is used, such as the Internet, mobile phone, social media, text messages, or apps.

Digital Supply Network

The digital transformation of the supply chain management industry is known as the Digital Supply Network. It strives to maximize value all along the manufacturer's supply chain, sales and distribution partners, and the end customer.

The Digital Supply Network enables manufacturers to receive raw parts and materials at the right time, in the right quantity, and the right order. Visibility into the supply chain allows manufacturers to reduce large and expensive inventories and follow Lean Manufacturing principles.

The Digital Supply Network also enables more responsive communication with customers, who demand fast deliveries and a positive Customer Journey.

Digital Twin

A Digital Twin is a replica of a physical machine or system where data flow is both to and from the physical object and the digital object and is more than just a reflection of the physical object. Digital Twin contrasts with the Digital Shadow, which has data flow from the physical object to the digital object only.

Digital Twins use data from sensors installed on physical objects to represent their near real-time status, working condition, or position. See Digital Shadow.

Digitization

The move to Digitization has allowed Industry 4.0 and the connected factory to develop, facilitating a fully integrated supply chain from a product's development right through to its final distribution. With all information being in an electronic format that a computer can understand, systems and machines can provide highly efficient operations.

In Industry 4.0, people, machines, and industrial processes are network-based on Cyber-Physical Systems incorporating state-of-the-art information and communications technology.

Distributed Control System (DCS)

DCS systems provide industrial control for process manufacturing applications, such as oil and gas production and refinement, chemical processing, and water treatment.

The applications usually have many analog inputs and outputs. They require advanced calculations, such as Proportional-Integral-Derivative (PID) to control the process.

See Programmable Logic Control (PLC) for a similar type of control concept but used in discrete manufacturing applications, such as machine control.

Edge Computing

Edge Computing reduces the amount of data sent to the cloud because data processing occurs at (or near) the sensor or data source's physical location instead of in the cloud. Most people consider the location of Edge Computing to be in the operational technology area of an enterprise.

Edge Computing provides faster response and reduces costs associated with transmitting data, especially Big Data, outside the facility.

An Edge gateway serves as a network entry point for operational technology devices to communicate with IT systems or a cloud. They also often provide network translation between networks that use different protocols; MQTT is often used.

See Fog Computing.

Enterprise Resource Planning (ERP Software)

Enterprise Resource Planning (ERP) refers to business-management software, usually comprising a suite of integrated applications.

Its job is to collect, store, manage, and interpret data from its many business activities. It includes inventory management, cash flow, and raw materials, CRM (Customer Relationship Management), production capacity, order status, purchase orders, and payroll.

ERP software allows individuals to access data from other departments using the same information to base their decisions upon a single source of information.

ERP software, in 2021, accounts for over 50% of the average manufacturer's software spend.

Flexible Manufacturing

Smart Manufacturing is also Flexible Manufacturing, which means that the factory uses modern equipment and processes to produce customized and personalized products. Short production runs resulting in Batch Size One, require machines to quickly re-configure, with short setup times for each job.

Fog Computing

Like Edge Computing, Fog Computing reduces data transfer to the cloud for analysis, improving security, processing speed, and cost. This creates efficiencies and has opportunities for companies concerned with compliance issues.

Typically Fog Computing is done in the local area network rather than in the sensor or gateway of an Edge Computing system.

Global Lighthouse Network

The Global Lighthouse Network is a World Economic Forum research project, in partnership with McKinsey & Company, that identifies leading Smart Manufacturing companies globally. As of 2021, 44 manufacturing sites are recognized as Industry 4.0 leaders and successfully scaled beyond the initial pilot stage. The six Key Performance Indicators for measuring performance are productivity, sustainability, agility, speed to market, and customization.

Hadoop

Administered by the Apache Software Foundation, Hadoop is open-source software that enables the distributed processing of large data sets across clusters of computers. It provides a framework for processing Big Data sets using specific software models, and specific use cases include quality control analysis, price planning, Predictive Maintenance applications, and many others.

HANA

HANA (High-performance Analytic Appliance) is a database created by the company, SAP. It processes high-volume transactions and provides real-time analytics of Structured Data. It is deployable on the customer premises, in a private or public cloud, and various hybrid configurations.

HART

HART, which stands for Highway Addressable Remote Transducer, is a protocol used by process instrumentation to send and receive configuration, diagnostics, and process data. HART data is transmitted by superimposing digital data on top of the standard 4-20mA signal loop, thereby allowing bidirectional communication with just a single pair of wires. Wireless HART is a version of HART that doesn't require communication wires to function.

Heijunka

Heijunka is a Lean Manufacturing method that aims to smooth or level the volume and mix of products the factory is manufacturing or assembling during each production cycle. This creates an environment that reduces the possibility for errors or defects to be made.

Horizontal Integration

This term denotes owning or controlling, different companies that are manufacturing similar products. An example would be a manufacturer of widgets that purchases several other companies making the same widget, thereby increasing market share.

Horizontal Integration also occurs within the same organization. There it relates to hierarchy levels and departments which carry out similar tasks. An example might be supply chain management, where the incoming materials, work in progress, and outgoing materials are tightly controlled and coupled.

One of its functions is to achieve optimized material and information flows. See Vertical Integration and Addendum III for a graphical representation of horizontal and vertical integration.

Human-Machine Interface (HMI)

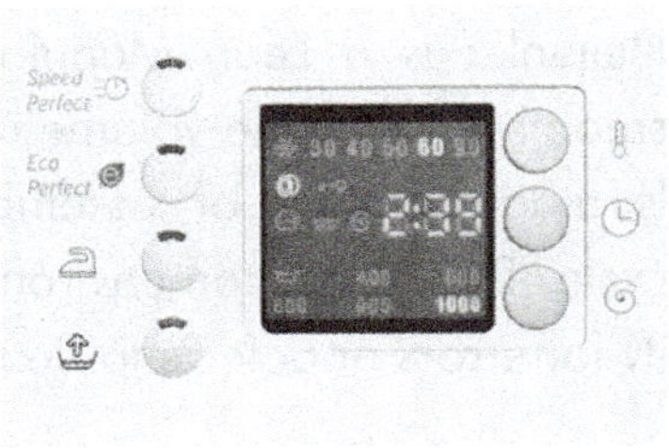

The HMI screen, or monitor, provides a graphical visualization of the industrial control and monitoring system. Previously called an "MMI" (Man-Machine Interface), an HMI typically displays information in theplant associated with either a Programmable Automation Controller (PAC),a Programmable Logic Controller (PLC), or a Distributed Control System (DCS).

Initially, HMIs replaced manual push-button interfaces so machine operators could interact with the controls. The HMI screen can also display information about the process, or operation, and machine health diagnostics.

The HMI will continue to have an increasingly important role in the Smart Factory as the need for information transparency and visibility continues to grow and involve employees of all job functions in the organization.

Hyperautomation

Hyperautomation is a term created by the company, Gartner, to describe the process of all technologies, but especially Artificial Intelligence and Machine Learning, to identify and automate as many processes as possible. It is also associated with creating a Digital Twin.

Industrial Ethernet

Industrial Ethernet refers to the use of Ethernet in manufacturing and other industrial environments that require modification of the"commercial grade" equipment used in offices and buildings.

Industrial environmental locations are harsher than commercial areas in terms of temperature swings, particular matter, vibration and shock, and electrical power quality issues. They require different physical layer devices, including special connectors, switches and hubs, and cables.

Industrial applications are often highly dependent upon fast andconsistent response times because moving equipment is synchronized. Low latency, low jitter, and high determinism related to response times are required compared to standard Ethernet.

Popular protocols of Industrial Ethernet include EthernetIP, Profinet, EtherCAT, SERCOS, and Modbus TCP.

Industrial vs Commercial Ethernet

	Industrial	Commercial
Connectors	M12, M23, Rectangular	RJ45
Environmental Rating	IP Classification	MICE Tables
Cable Requirements	Oil Resistant, High Flex	Plenum or Riser Rated

Industrial Internet Consortium

The Industrial Internet Consortium, a non-profit consortium, was founded in 2014 to accelerate the adoption of the Industrial Internet. It publishes the Industrial Internet Reference Architecture (IIRA) Technical Report.

It promotes best practices for commercial use, and its members include both large and small companies, researchers, and government organizations. Key industry sectors include manufacturing, mining, transportation, energy, healthcare, buildings, utility infrastructure, defense, and transportation.

Founding members include Dell, EMC, General Electric, Huawei, Microsoft, and Purdue University.

Industrial Internet of Things (IIoT)

A system that connects and integrates industrial control systems with enterprise systems, business processes, and analytics using Internet of Things (IoT) technologies in an industrial setting.

Just as Industrial Ethernet requires more rugged and specialized products, IIoT devices need robust construction to handle the environmental conditions of factories, plants, and utilities.

IIoT refers to the broad description of devices and systems used in industrial environments, whereas Industry 4.0 focuses on manufacturing. See Addendum I for a graphic illustrating the relationship between IoT, IIoT, Industry 4.0, and Advanced Manufacturing.

Industrial Internet Reference Archit ecture

This standards-based architectural template and methodology enable Industrial Internet of Things (IIoT) system architects to design their systems based on a common framework and concepts published by the Industrial Internet Consortium.

Industry 4.0

Industry 4.0 is the common name used to describe the current trend towards a fully connected and automated manufacturing system, which is also known as the Smart Factory or Smart Manufacturing. Spelled Industrie 4.0, its announcement was made at the Hanover Fair in 2011.

Developed in Germany, it postulates that we are currently experiencing the Fourth Industrial Revolution, which uses cyber-physical systems to manufacture small batches of personalized products.

The current trend in the manufacturing industry uses a combination of IoT, big data, and cloud computing to develop factories that can make decisions based on large amounts of data. See Addendum I for a graphic illustrating the relationship between IoT, IIoT, Industry 4.0, and Advanced Manufacturing.

Internet of Things (IoT)

IoT refers to a network of physical objects connecting and sharing information with other devices via the Internet. Many consumer devices such as smart speakers, doorbells, toasters, watches, thermosets are part of the IoT.

Reprinted with permission from "The Smart Student's Guide to Smart Manufacturing and Industry," see Addendum I for a graphic illustrating how IoT, IIoT, and Industry 4.0 are related.

An IoT gateway connects physical devices to a Cloud Computing platform.

Interoperability

Interoperability is the ability of systems, designed by different parties, to communicate with each other. Standards allow devices from different manufacturers to work together, such as a Bluetooth headphone set from one manufacturer connected to a smartphone from a different one.

Web Services, MQTT, and Application Programming Interfaces (APIs) are other popular methods of achieving interoperability in IIoT and Smart Manufacturing operations.

IO-Link

IO-Link is an open communication standard (IEC 61131-9) that defines sensor and actuator communication with a controller, typically a Programmable Logic Controller (PLC). Automation applications, especially in discrete manufacturing processes, use IO-Link, whereas process manufacturing applications may use HART.

JavaScript Object Notation (JSON)

JSON is an open-standard file format that uses human-readable text to communicate information. See AMQT, MQTT, and XML for similar concepts.

Jidoka

Jidoka is a Lean Manufacturing technique aimed at preventing inventory building up if one part of the machine or factory has a problem and finds a permanent solution for the issue in an urgent manner.

Just in Time

Just in Time is a Lean Manufacturing method aimed at increasing return on investment by reducing inventory levels to a minimum while maintaining reasonable delivery times to end-users. It requires efficient management of the supply chain in order to avoid disruption.

Kaizen

Kaizen is a Lean Manufacturing method to improve work processes continuously. In Japanese, it means "change for the better" or continuousimprovement.

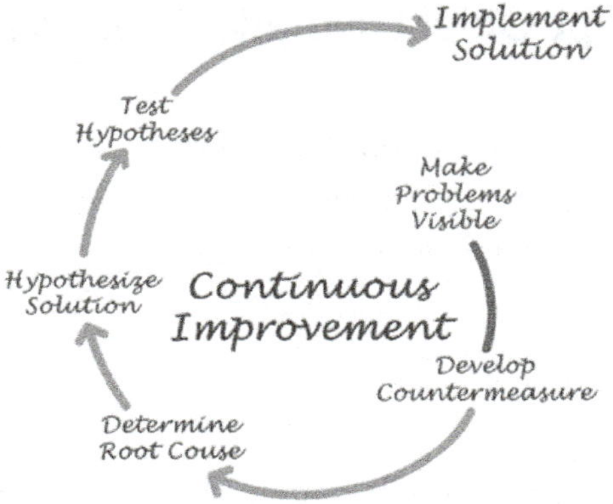

Kanban

Kanban is a Lean Manufacturing method to avoid excess inventory in the production line, especially by minimizing work-in-progress. Kanban meansa sign or billboard, and it uses cards to visualize the work in progress. As tasks are finished, the cards are moved accordingly.

Key Performance Indicators (KPI)

KPIs are quantifiable measurements to evaluate the performance of industrial solutions and processes within an operation. KPIs can be benchmarked and compared across multiple facilities and include inputs spanning the number of deliveries made, average downtime, cost per labor per shift, etc.

A Key Performance Indicator is a measurable value that demonstrates how effectively a company is achieving key business objectives. Organizations use KPIs to evaluate their success at reaching targets.

Lean Manufacturing

Lean Manufacturing is a quality system and manufacturing philosophy that ensures an efficient flow of materials, components, and finished products.

Toyota coined the concept to ensure manufacturing operations ran as efficiently or as leanly as possible. Lean Manufacturing identifies *mudas* (or wastes) as a critical practice to avoid in manufacturing. For more details on some key Lean Manufacturing techniques, see Heijunka, Takt-time, Jidoka, Just in Time, KANBAN, 5S, Poka-Yoke, and Kaizen.

Lot Size One

See Batch Size One.

Manufacturing Cell

A manufacturing cell arranges the equipment necessary to create a production for a single workspace. This contrasts with more traditional systems where similar machines, like drill presses, are located together.

Often organized in a u-shape, the cell facilitates shorter production times and increased efficiencies between each step.

Creating a manufacturing cell is generally more expensive in terms of capital equipment expenditures because duplicated equipment is locatedin different plant areas. However, there are several ongoing cost savings from reducing inventory and waste that offset the capital costs.

See Chaku Chaku for a related concept.

Machine -to-Machine Communication (M2M)

Machine-to-Machine communication is the equipment's ability to interact with others without requiring human assistance. M2M can involve remote machine maintenance and monitoring machines, automatic machines, such as mobile pay terminals, or mobile data transfer. M2M brings together information and communication technology. See Interoperability,

Machine Learning (ML)

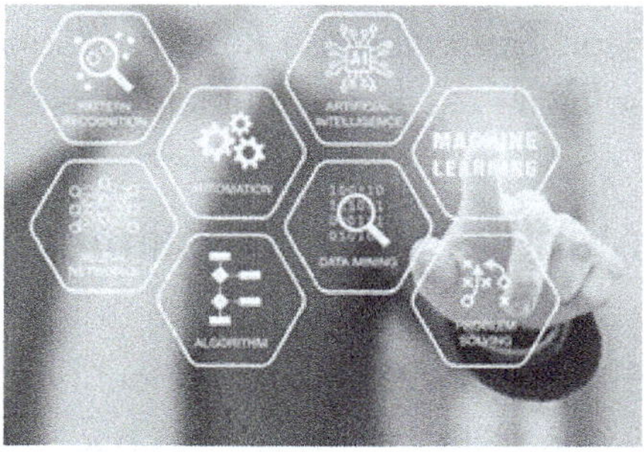

Machine Learning is a specific type of Artificial Intelligence and is how devices gain their "knowledge."

Large amounts of data are processed to recognize patterns, identify correlations, and apply rules. They can then detect anomalies and react accordingly. The learning process may occur within an individual device or connected to an intelligent network. Learning occurs autonomously without explicit programming from humans.

In Industry 4.0, Machine Learning is effective in optimizing production processes, according to real-time information, within rules or constraints. See Deep Learning and Artificial Intelligence and Addendum IV

Manufacturing Execution System (MES)

Manufacturing Execution Systems are production software systems that provide management and control of manufacturing operations in real-time.

MES provides information that helps manufacturing decision-makers understand how to optimize plant floor operations to improve production output. MES works in real-time to enable the control of multiple production process elements (e.g., inputs, personnel, machines, and support services).

MES, in 2021, is the fastest-growing software spend by manufacturers, with an estimate of over 13% growth every year. See Enterprise Resource Planning for another popular software used in manufacturing.

Manufacturing Institutes

The United States government funds a network of Manufacturing Institutes to develop and commercialize manufacturing technologies within the United States. The Department of Defense is concerned about its supply chain's stability when many components needed formilitary operations only have foreign sources of supply.

The list of Institutes operating in 2021 include:

- Advanced Functional Fabrics of America, Cambridge MA

- American Institute for Manufacturing Integrated Photonics, Rochester NY
- America Makes, Youngstown OH
- Advanced Robotics for Manufacturing, Pittsburgh PA
- BioFabUSA, Manchester NH
- Bioindustrial Manufacturing and Design Ecosystem, Saint Paul MN
- Clean Energy Smart Manufacturing Innovation Institute, Los Angeles CA
- Cybersecurity Manufacturing Innovation Institute, San Antonio TX
- Institute for Advanced Composites Manufacturing Innovation, Knoxville TN
- Lightweight Materials Manufacturing Technologies LIFT, Detroit MI
- MxD Manufacturing time Digital, Chicago IL
- NextFlex, San Jose CA
- The National Institute for Innovation in Manufactur-ingBiopharmaceuticals, Newark DE
- Power America, Raleigh NC
- Rapid Advancement in Process Intensification De-ploymentInstitute, New York NY

- REMADE Reducing Embodied-energy and Decreasing Emissions, Rochester NY

Mass Customization

One goal of Industry 4.0 is to create one-off, consumer-defined productsin a cost-effective way, using flexible manufacturing processes and configurations. In contrast to mass production lines that produce identical products, mass customization makes personalized products by incorporating more sophisticated automation and controls. Product personalization requires Flexible Manufacturing to be able to changemachine configurations quickly and efficiently for every order. See Batch Size One.

Milkrun

Milkrun is a manufacturing logistics term referring to the material flow within a factory. The concept is part of Lean Manufacturing to keepinventory levels down and based on 'pull' production methods of replacing materials regularly. Milkruns work better when the facility produces highly similar parts or items requiring the same sub- components.

MindSphere

Mindsphere is a cloud-based IoT system offered by the international company, Siemens, that runs on Azure, AWS, and other platforms.

Mixed -Model Production

Mixed-Model Production is the manufacturing of several different product models within a Smart Factory, without requiring machine changeover. The goal is to provide optimal input to later workstations, for improved overall productivity, smooth demand on suppliers, and reduce inventory.

Mobile Robotics

Mobile robotics can be described as a combination of Automated Guided Vehicle (AGV) and Collaborative Robots or Cobots. Mobile robots may use sensors to autonomously determine their path rather, than following a predefined one that may not be optimal, or possible, at any onetime.

MQTT

The acronym stands for Message Queuing Telemetry Transport, and it is a lightweight, open, and simple messaging protocol, designed for device connectivity, especially for IoT applications. Invented in 1999, by IBM and others, it controlled a Conoco Phillips pipeline within a SCADA network.

MQTT became an open communication protocol standard by the Organization for the Advancement of Structured Information Standards (OASIS), in 2014.

It is an optimal solution for Machine-to-Machine Communication and IoT implementations, where a small code footprint is desired and limited network bandwidth is available. MQTT consists of clients that publish and subscribe to each other's messages through another participant, called the broker.

National Center for Defense Manufacturing and Machining

Organization founded to revitalize the US manufacturing industry to cut costs, increase productivity, improve quality, and reduce waste associated with products manufactured for national defense applications.

No-Code and Low-Code

Rapidly build customized computer code for applications and enterprise software platforms with little or no programming required. Use of graphical programming with drag and drop functionality. Some possibilities are quality and safety reporting apps, energy calculation, and maintenance. Several platforms are already available from companies such as Siemens, Microsoft, Oracle, Google, Appian, and others. They all allow people with domain knowledge of any discipline to create tools and apps without learning computer programming.

Node -RED

Node-RED is an open-source programming software, released in 2013, that provides a way to connect devices and APIs with flow-based programming. It is part of the No-Code and Low-Code movement.

OSI Model (Open Systems Interconnection Model)

OSI is a standardized model that describes how different network components communicate with each other. The model divides this communication into seven layers, also called stations.

For example, between sender and recipient, an e-mail goes through all these standard stations and arrives only if each layer/station fulfills its task.

The respective layer receives the necessary information on the object (sender/recipient, file properties) and carries out its function accordingly. All links within a process must know the procedure, irrespective of whether sender and recipient use different components and software.

1	*Application*
2	*Presentation*
3	*Session*
4	*Transport*
5	*Network*
6	*Data Link*
7	*Physical*

OSI Model

The OSI model describes how different components in a network communicate with each other. OSI divides communication into seven layers, called stations. All stations are recorded in a protocol to receive the necessary information from the inner layer and thus function effectively.

OPC UA (OPC Unified Architecture)

OPC UA, or *OPC Unified Architecture*, was developed by the OPC Foundation, an industry consortium focused on automation, and released in 2008, as platform independent service-oriented architecture. OPC UA refers to standards set for automation equipment in Machine-to-Machine communications.

From industry expert, Garrett Schmitt: "Unlike other market-leading protocols, like Ethernet/IP, Profinet, and Modbus TCP, OPC UA has no ties to a single vendor. The standard was created by the OPC Foundation, a consortium of member companies who are committed to ensuring the increasingly high standard for secure, reliable, and platform-independent information exchange. Since it is not linked to a specific manufacturer, it encourages vendors to create innovative solutions without competitive barriers and gives users the freedom to choose the best-in-class devices for their application."

There are two methods of exchanging data with OPC UA: a client-server model with HPPS/TCP and a publish-subscribe model with AMQP or MQTT. The Azure and AWS clouds support OPC UA, allowing industrial machines to be connected.

Operational Technology

Operational Technology is the hardware and software that monitors and controls physical devices, processes, and events in the enterprise. It describes plant floor technology and equipment used in production similar to the Information Technology (IT) used to run ERP programs.

OT may be defined as "IT + Physics," alluding to the fact that OT controls moving equipment and processes, not just displaying graphics or reports.

Increasingly, individual pieces of OT equipment need to communicate seamlessly with each other, requiring conformance to applicable standards. The fact that OT is incorporating more devices that originated from IT and IoT technologies, the Attack Surface and Vectors are increasing the necessity for robust Cyber Security measures.

OT/IT Convergence

OT/IT Convergence is the joining of Information Technologies and Operational Technologies. Operational technology is the equipment needed to make the factory run and produce the product.

Convergence describes the OT control of assets in the Smart Factory in real-time, with IT's communication and data analysis power. Edge Computing devices help create convergence by processing real-time information and making decisions based upon it.

Overall Equipment Effectiveness (OEE)

OEE is the evaluation of how effectively equipment is working in a manufacturing environment.

OEE is one of the most widely used standards for improving manufacturing productivity. OEE provides a simple score of how well a factory or specific piece of equipment is producing compared to the best that it could be. It achieves this by calculating how close to perfect the parts are (Quality), how many are made (Performance), and without interruption (Availability).

For example, a factory that achieves 90% each for Quality, Performance, and Availability would have an OEE score of 72.9% out of a possible 100%.

Often OEE is used as a Key Performance Indicator (KPI) and often used in Lean Manufacturing.

OEE Calculations

$$OEE = A \times P \times Q$$

A = Run Time ÷ Planned Time where

Run Time = Planned Time - Stop Time

P= Total Count ÷ Run Time ÷ Ideal Run Rate

Q= Good Count ÷ Total Count

Performance and Performance Loss

Performance is how close the equipment is to producing good parts atthe design speed. Performance is one of the three factors of Overall Equipment Effectiveness or OEE.

Performance Loss occurs when the equipment or line operates at slower than optimal speeds or stops and interrupts production.

Plattform Industrie 4.0

Plattform Industrie 4.0 is a German group to promote Industry 4.0 that cooperates with the American-based Industrial Internet Consortium and other international organizations.

Poka-Yoke

Part of Lean Manufacturing aimed at preventing mistakes by developinga system of fool-proof processes that prevent errors from occurring and improve the quality of the end product.

Predictive Analytics

Predictive Analytics is the practice of extracting information from existing data sets to determine patterns and predict future outcomes and trends. Using predictive analytics, we can predict future events, like when a machine might fail. It is a vital part of enabling Predictive Maintenance programs.

Predictive Maintenance (PdM)

Predictive Maintenance (PdM) aims to use the data collected from sensors and connected machines (M2M) in the Digital Manufacturing environment to predict when a failure is about to occur and order proactive maintenance to prevent it from happening. The goal of a PdM program is to maximize the utilization of equipment by minimizing downtime.

PdM is in contrast to Preventative Maintenance that orders maintenance based on a time interval. This maintenance is expensive because it authorizes work based on worst-case averages, not when the maintenance is actually required.

Machine Learning can be applied to the data-to-spot trends and patterns, specifically for maintenance, avoiding unscheduled downtime, and maximizing overall machine availability, resultantly improving OEE scores.

Predictive maintenance is also being extended into the field after product delivery, to facilitate Servitization, where the supplier performs additional services to generate revenue.

Predictive maintenance used in a Servitization model has the potential for machine manufacturers to receive data from equipment installed at the customer's factory. Understanding how a product is used and detecting defects (and ideally, remote maintenance) leads to better future design and improved customer relationships.

Predictive Maintenance uses sensors and other data to detect abnormal conditions, thus allowing Predictive Maintenance in a planned manner.

Preventative Maintenance

Preventative Maintenance was considered the best way to prevent unplanned downtime of equipment by periodically performing maintenance at set intervals.

A classic example is changing automotive motor oil every 5000 miles or six months. While effective, it has drawbacks, including the expense of replacing those that don't need to be replaced, accidentally breaking those things, and having equipment out of service during that time.

The development of more advanced sensors and data analysis are pushing organizations towards Predictive Maintenance instead.

Production Analytics

Production Analytics software is a platform that helps you to visualize, in real-time, the data of your machines, lines, and factories. In addition tolive tracking, Production Analytics review historical data and provide a visualization of production losses to realize solutions.

Programmable Logic Controller (PLC)

Programmable Logic Controllers are electronic controllers that provide a logical sequence of events commonly found in discrete manufacturing processes and applications. They offer a fast response of inputs and outputs and traditionally incorporate and analyze many binary or discrete signals (rather than analog signals) and work in a type of IF-THEN manner.

PLCs are programmed with a software programming tool. The most common one used in North America is Ladder Logic. In other parts of the world, PLC programming conforms to the IEC61131 standard, which provides additional languages, including Structured Text, Instruction List, Sequential Function Chart, and Function Block Diagram.

Distributed Control Systems (DCS) are more often used in process industries and applications because of the need to analyze more analog signals and perform advanced calculations.

Quality and Quality Loss

Quality is how close to perfect the part run is, with 100% being the maximum. Availability is one of the three factors of Overall Equipment Effectiveness or OEE.

Quality Loss includes lower product yield than expected or defects in the product.

Radio Frequency Identification (RFID)

RFID devices are chips that communicate with a reading device using an electromagnetic field. RFID semiconductor chips, or tags, are attached to an object and are read with scanners to identify the item.

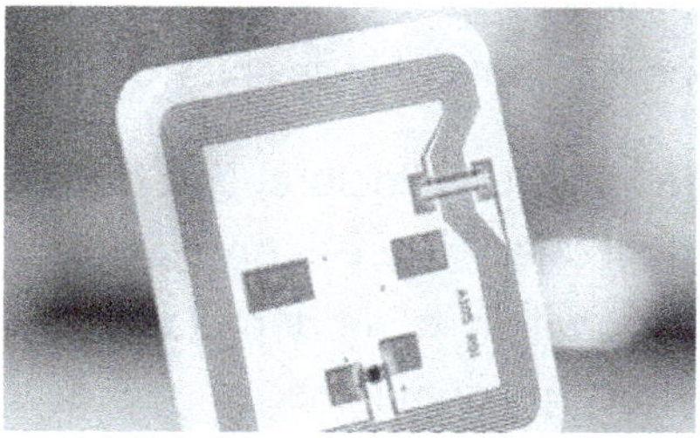

Applications in manufacturing include component tracking of parts in inventory and during the production sequence. Information can be written and read from the chip many times, providing up-to-date status. In Smart Manufacturing, information about the customer's order is written to the RFID chip, so the appropriate actions are taken during the manufacturing processes.

Another use is to provide access control to equipment controls by identifying the operator who is accessing the machine.

Shown below is one type of RFID chip with an integrated antenna forming concentric circles.

RAMI 4.0 (Reference Architecture Model I 4.0)

RAMI 4.0, referred to as the "map" of Industry 4.0 solutions for many different disciplines, was developed by ZVEI, the German Electrical and Electronic Manufacturers' Association, to support Industry 4.0 initiatives. It is a three-dimensional structural model that presents all levels and participants of Industry 4.0 in an understandable manner.

In this architecture, processes are divided into smaller packages. There is an axis for the grid-like hierarchy structure within a modern factory, an axis for the architecture structure (functions, processes, data), and a third axis that describes the product life cycle. It is intended that RAMI 4.0 functions on a global basis, identifying and bringing together all developments in the manufacturing industry, securing a standardizedexchange of information.

RAMI 4.0 combines all elements and IT components in a layer and life cycle model. RAMI 4.0 breaks down complex processes into easy-to-grasp packages, including data privacy and IT security.

Re-Shoring

Reshoring refers to the reversal of the massive "off-shoring" movement of the 1990s, primarily in the United States, that resulted in manufacturing operations moving to other countries with lower labor rates.

The 2019 COVID crises and geopolitical pressures have made the re- shoring of manufacturing a matter of national security and safety.

Industry 4.0 and Smart Factory technologies make reshoring possible.

An advantage of Re-Shoring is that a product is readily available and with a high level of customization or personalization, with less risk of supply chain interruptions.

Due to the high degree of automation, manufacturing currently outsourced to low-wage countries will shift to locations close to the customers because intelligent automation allows cost-efficient and high-quality production.

Risk-based Planning & Scheduling (RPS)

Risk-based Planning and Scheduling extends traditional planning andscheduling techniques to fully account for the variation present in nearly any production system.

Using a simulation model, RPS generates a detailed resource-constraineddeterministic schedule and a probability-based risk analysis of that schedule to account for variation in the system. In this way, RPS is used to generate plans that minimize risks and reduce costs in the presence ofuncertainty.

An example of an application area of Industry 4.0 is remote maintenance that permits maintaining a large printing machine at a customer's location in Australia, from a supplier based in Germany.

When combined with Augmented Reality, this becomes even more possible. For this, the installer in the field would put on special glasses(for instance, Google Glass) to identify the exact screws that need to be removed in the correct sequence, to repair the machine, all via visual displays on the glasses.

SCADA (Supervisory Control and Data Acquisition)

SCADA refers to a control system that monitors widely dispersed assets in electric power generation and distribution, water treatment, and oil and gas operations. In the past, SCADA was tied heavily to the field of "telemetry", which is the remote and automatic transmission of information over large distances.

Historically, popular vendors were General Electric, Westinghouse, and Leeds & Northrup, where the author of this book worked for a short time in the 1990s.

Today the term often refers to any display system that shows an overview of an entire system or manufacturing plant or production line.

The following is from industry expert Christine Frank:

SCADA is used either "plant-wide" or at the "machine/asset/equipment level." For instance, SCADA used in the process - is plant-wide, site-wide, or field-wide to give visibility to large-scale deployments. These applications are typically connected to a DCS or PAC with 1000's of I/O (process instrumentation), and usually are managing continuous movement of liquids, gas, fibers -- things that can't be taken apart during the process or stopped midstream, without being managed during the entire flow of steps,

Whereas "Discrete automation", HMI/SCADA is used on the machine cell level to manage functions and recipes on a piece of equipment or asset -- including batch processing, assembly, CNC's, machine tools, robotics, production applications in factories, and manufacturing facility, etc.

Whereas "building automation systems", SCADA/HMI's monitor and control the boilers, furnace, rooftop handlers, compressors, heat pumps, etc., in order to manage the functions and recipes of the heating/cooling system, via an HMI panel that uses SCADA functionality.

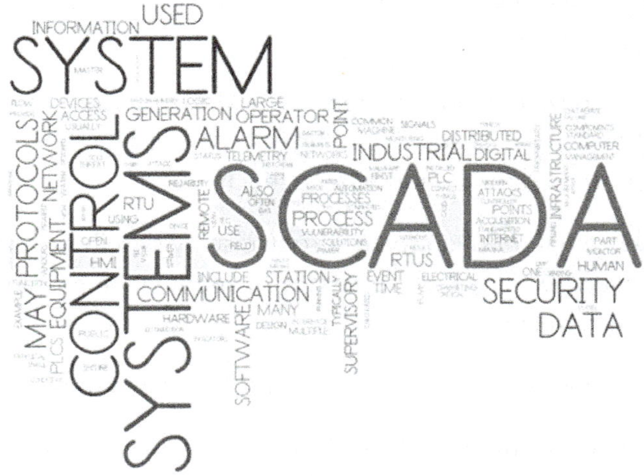

Servitization

The idea that a manufactured product can generate increased value afterits purchase, by bundling services for the customer and guaranteeing desired outcomes. Customers are looking for results; manufacturers are looking for reoccurring revenue streams, so there is a desire for more of this type of business arrangement on both sides. See XaaS for related concepts.

SMAC

SMAC was a popular acronym around 2010 and stood for "Social, Mobile, Analytics, Cloud." Working together, these four technologies link products to their end-users in the manufacturing context, providing instant feedback into market trends and preferences, which drives innovation and growth. In the author's opinion, SMAC hasn't been used much from 2015 on, as it was superseded by IoT.

Smart Device

Examples of Smart Devices include modern mobile phones and tablets. Smart Devices connect the Internet or other devices via Bluetooth, NFC, Wi-Fi, 5G, and communication methods. Smart Devices may be thought of as a combination of a small computer with a communication channel.

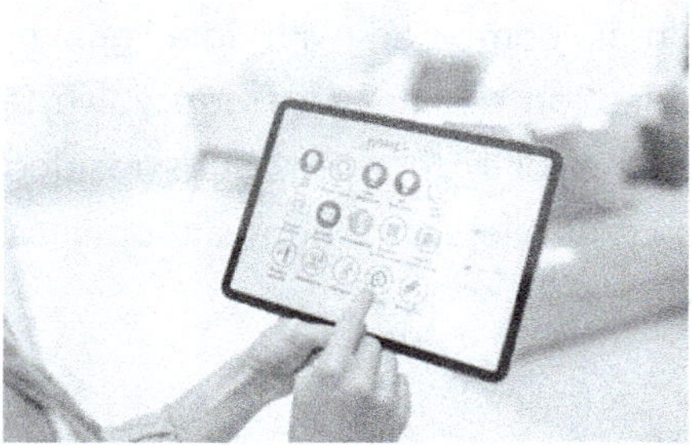

Smart Factory

It is an industrial environment that uses Industry 4.0 technologies, such as cloud computing and networked activity to offer task completion through autonomous capability. Information about all activity within the plant is received through real-time data reporting, which can be used to adjust issues before they arise.

Smart Manufacturing

An environment where computers control decision-making in which the physical and digital are connected and communicate with one another to improve production. Mostly synonymous with Smart Factory and Industry 4.0.

Smart Products

Consumer Smart Products include the ability to access the cloud fordownloading updates and data. A smart speaker or doorbell camera is an example of this.

Industrial smart products are similar in that the product's base function is combined with intelligence, networking, and communication abilities. A smart valve terminal might include logic capability that previously resided in Programmable Logic Controllers or diagnostic abilities to assess its own condition.

Soft Skills

Smart Manufacturing requires highly developed people skills for all humans involved in the organization.

As humans give robotic and repetitive tasks (both physical and mental) to machines and automation, the jobs that remain will require workers to function well in teams and individually.

Soft Skills now mean more than just obtaining proficiency in reading, writing, time-management, and verbal communications.

Over Time more skills are being added to the list of requirements for success in the Smart Manufacturing workplace, including emotional awareness and cultural sensitivity.

The graph below from "**The Smart Student's Guide to Smart Manufacturing**" shows many essential soft skills required in the Industry 4.0:

Software as a Service (SaaS)

Software as a Service (SaaS) is a form of software distribution and marketing. In the SaaS model, the software supplier is responsible for all the necessary structures to make the system available (servers, connectivity, information security care). The client uses the software via Internet, paying a fee for the service.

The SaaS model offers software as a service with specific purposes available to users on the Internet. Software systems are accessible from multiple devices, through a client interface, on a client-server model network, such as a Web browser.

In SaaS, the user does not manage the application's features, except for specific configurations. Therefore, developers focus on updating rather than infrastructure, leading to the rapid development of software systems.

Software is delivered to the user and updated via the Internet, and typically enabled by a cloud service that hosts the software. Often there is no software installed on the user's device, but the information is accessed via a web browser.

This business model is being expanded to include many other things sold as a service. See Servitization and XaaS for more information.

Structured Data

Structured Data presents its information in an orderly manner. A spreadsheet with accurate data labels and data formatted consistently is an example of Structured Data. In contrast, Unstructured Data can consist of free-form text, images, videos, social media feeds, audio files, and more. See Data Attributes, Variety.

Subtractive Manufacturing

Subtractive manufacturing creates products by removing material in a controlled manner, usually with a machine (See Computer Numerical Control (CNC)). Typical operations include turning, milling, laser cutting, wire EDM (Electrical Discharge Machining). For a different type of manufacturing that is becoming increasingly popular, see Additive Manufacturing and 3D Printing.

Takt-Time

Takt-Time is a Lean Manufacturing technique to prevent excess inventory by balancing the various processes, upstream and downstream. It is the measure of time between the production start of one unit to the next production start.

TCP/IP

Transmission Control Protocol/Internet Protocol specifies how Internet data is packetized, addressed, and transmitted. It was initially referred to as the Department of Defense model because its funding came from the US Defense Advanced Research Projects Agency. The four layers are Application, Transport, Network, and Physical.

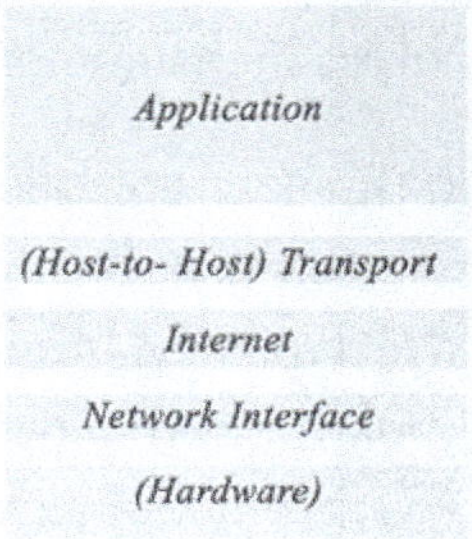

TCP/IP Model

Traceability

Traceability refers to the ability to fully identify the path all raw materials, producers, upstream suppliers, and individual parts take from manufacturer to customer. It determines when, where, and by whom the goods were produced, processed, stored, transported, used, or discarded.

Traceability occurs in two directions: from the manufacturer to the consumer and from the consumer to the manufacturer. Traceability may use Blockchain technology to ensure data security and integrity.

Transmitter

A transmitter combines a sensor and communication system to measure and communicate process variables, such as flow, pressure, temperature, and level -- and specialized liquid and gas analysis measurements.

Transmitters are often mounted inside a weatherproof and explosion-proof metal housing, with or without a visual display. Typically, transmitters provide an analog 4-20 mA signal, proportional to the variable they are measuring. The HART protocol increases the amount of data communicated by adding a signal on top of the 4-20mA loop.

Very often, Transmitters provide the input signals to a Distributed Control System (DCS).

Unstructured Data

Information that either does not have a predefined data model or is not organized in a predefined manner. It cannot be stored in a database in structured fields.

Examples of Unstructured Data include text, images, audio, video, orTwitter feeds, to detect patterns or events that are explicitly labeled. Unstructured Data, in contrast to Structured Data, is more unorderly and challenging to analyze. See Data Attributes, Variety.

User experience (UX)

User Experience (UX) is the set of elements and factors related to user interaction with a given product, system, or service whose result generates a positive or negative perception. An essential concept in UX design is the process by which users form experiences. When the user encounters a product, it forms a momentary impression, which evolves. Creating a great UX requires a team approach with designers, engineers, and testers providing input to the final design. UX is part of theCustomer Journey.

Vertical Integration

Vertical Integration in manufacturing is a company's strategy to own or control parts of its supply or value chain. Such a company would bring product development, critical material supplies, and sales channels into its control.

Vertical Integration can be realized by buying companies that provide key technology to assure continued access to its products and to influence its product roadmap. It can also include downstream integration, where parts of the distribution or retail channels are controlled or owned by the manufacturer.

Vertical integration occurs within a single plant as well. An example is grouping all the data in a single solution as it moves from the factory floorto the control level, up to the MES and ERP. See Horizontal Integration and Addendum III for a graphic illustration.

Virtual Reality (VR)

Virtual Reality is a computer-generated simulation of three-dimensional space or image that interacts with humans in a seemingly natural way. Typically, a person uses goggles like the one shown below, which immerses the user into a computer-generated visual field. Increasingly other senses are also incorporated into the VR experience.

In the manufacturing environment, it can allow for rapid visualization, prototyping, and simulation. See Augmented Reality.

Web Service

An interface standard based on the HTTP protocol uses a request and response transaction for a web page. It is related to XML.

White Label

A white label is a product (or service) that is sold and marketed by another party. A contract manufacturer, hired to produce products under the customer's name, is an example of a white label.

Wi-Fi

Wi-Fi was a name chosen to be more appealing to the public than "IEEE 802.11b Direct Sequence" and has no direct meaning and is not an acronym. Wi-Fi is a trademark of the Wi-Fi Alliance, which restricts the term "Wi-Fi Certified" to products that complete interoperability certification testing.

Wi-Fi is a technology for wireless local area networking with devices based on the IEEE Ethernet standards.

Wi-Fi Generations

Standard	802.11a	802.11b	802.11g	802.11n	802.11ac	802.11ax
Name	Wi-Fi 1	Wi-Fi 2	Wi-Fi 3	Wi-Fi 4	Wi-Fi 5	Wi-Fi 6
Rate	54Mbps	11Mbps	54Mbps	600Mbps	1300Mbps	12Gbps
Year	1999	1999	2003	2009	2014	2019

©

XaaS (Everything as a Service)

Sometimes referred to as Anything as a Service, it converts businesses that were previously only involved in product sales, into subscription-based models, ensuring a reoccurring revenue stream. The idea is that customers value outcomes rather than features and benefits.

XML (Extensible Mark-up Language)

In computing, Extensible Mark-up Language is a text format language that defines rules for encoding documents in a human-readable and machine-readable way. It was developed by the World Wide Web Consortium, initially for addressing document formatting online. It is used to communicate data between applications, communication protocols, graphics, and other uses.

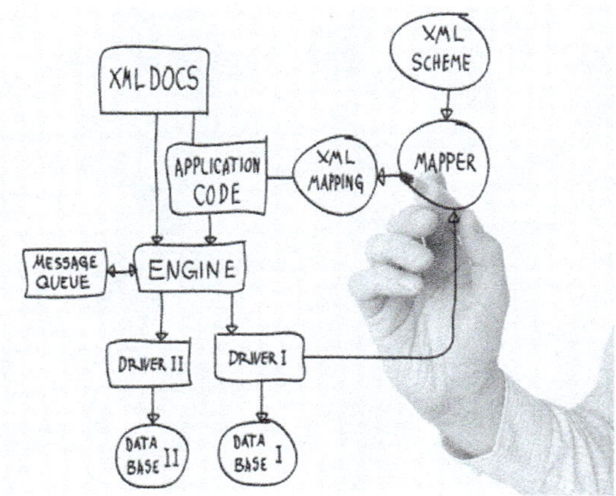

Yield Improvement

The improvement in good parts described as the percentage of decrease in bad parts. For example, if yield improves from 92% to 96%, the Yield Improvement is 50%, the difference between creating 8% bad parts down to 4%.

ZigBee

ZigBee is a low-power radio protocol for small amounts of data, based on the IEEE 802.15.4 standard, and published by the Zigbee Alliance formed in 2002. It has low power consumption, a range of about 100 meters, and a bandwidth of 250 kbps.

Addendum I - IoT, IIoT, and I4.0

The terminology around Smart Manufacturing is not strictly defined. Below is the author's interpretation.

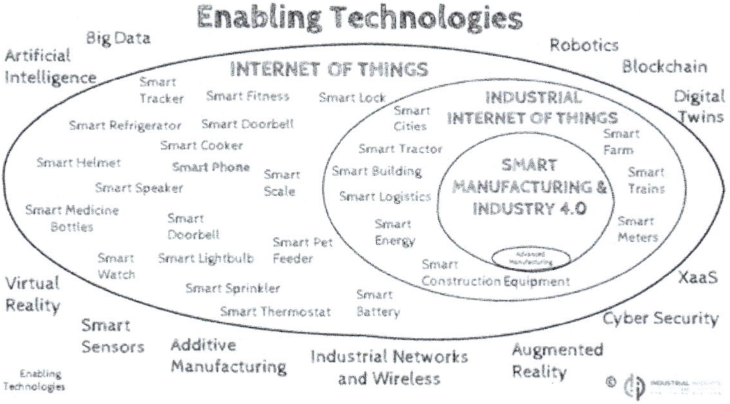

Addendum II –3D Printing Technologies

Several types of 3D printing technologies exist, with the most popular illustrated below:

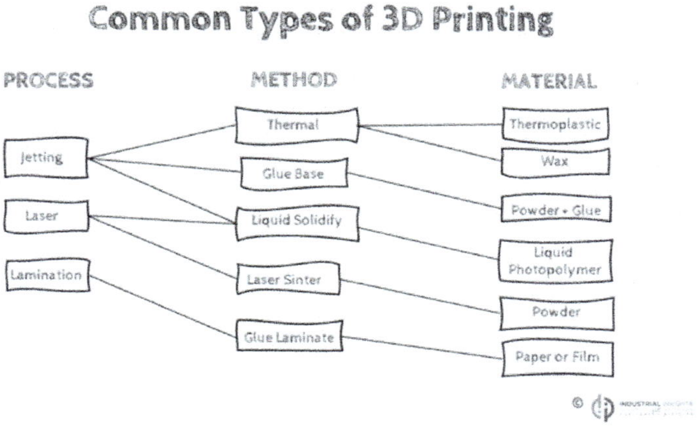

Addendum III – Integration Types

An example of Horizontal Integration is a car manufacturer purchasing ownership in other car manufacturers, creating greater scales of efficiency, and capturing market share.

An example of Vertical Integration is a car manufacturer that purchases or otherwise controls the companies *backward* in its supply chain, to have greater control over those raw materials, or forward, by owning or managing the manufactured item's sales channel.

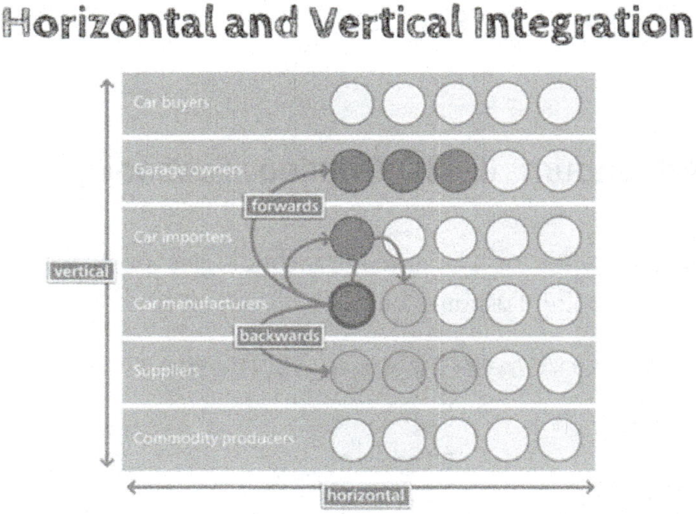

Derivative work of Martin Sauter derivative work: Andrew c (talk) - Integration.png

Addendum IV – Guide to AI Terminology

The terminology around artificial intelligence is used loosely in many texts but there are some differences.

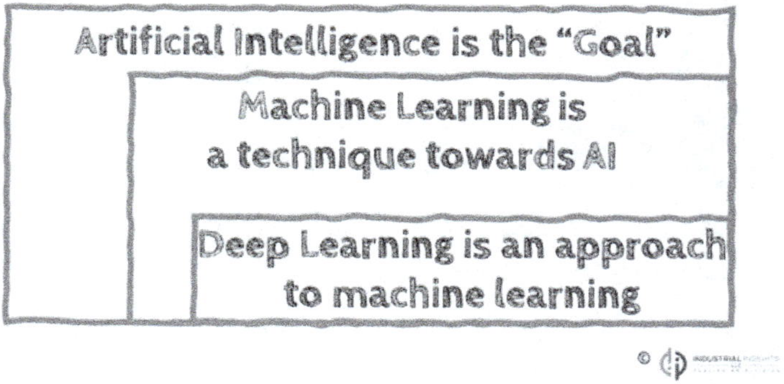

About the Author

Mike Nager is an engineer with more than 20 years of experience working for leading industrial manufacturers. He has consulted with hundreds of manufacturers and utilities to design more efficient, environmentally friendly, responsive plants and factories.

He was named a 2020 Top 10 IIoT Influencer by Onalytica and haspublished dozens of trade publications.

He has installed sensors inside the massive machine tools used in making automotive engines, tested industrial radio technology deep underground in the New York subways, and wore a cleanroom suit to work with clients onsite in both the pharmaceutical and semiconductor industries.

In 2018, at the Nevada Economic Development Conference, Mike was a keynote speaker, and in 2019 at the Réseau Innovation 4.0 Network in Quebec.

Mike is a volunteer with the IEEE Mini-Engineering Academy that provideshigh school students opportunities to work with engineers on hands-on projects in many areas. He is an advisory board member to educational institutions, including the Monmouth County Vocational School District.

He lives at the Jersey Shore with his fantastic wife, two daughters, and one ugly rescue dog.

Other Publications by Mike Nager:

The Smart Student's Guide to Smart Manufacturing and Industry 4.0, available in Ebook and Paperback

Made in the USA
Monee, IL
04 November 2026

67380080R00056